Sir Isaac Newton
Inventor of the Doggy Door

I0494632

By Gary "Two Horse" Green
and Aaron Brachfeld

VOLUME 3 of the TWO HORSES series

ISBN-13: 978-1530389971

ISBN-10: 1530389976

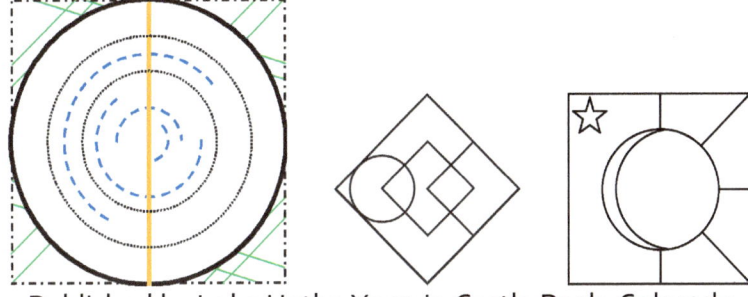

Published by Loka Hatha Yoga in Castle Rock, Colorado

There are many suitable places for your meditation.
We hope this becomes one of them.
Anguttara Nikaya 1.159

LOKAHATHAYOGA@GMAIL.COM - (719) 422-9536
http://lokareview.blogspot.com/p/books.html
lokahathayoga.blogspot.com

I can calculate the motion of heavenly bodies, but not the madness of people.

– Sir Isaac Newton

ABOUT THE AUTHOR

Gary Two Horse Green is a husband, father, and veteran of the Marine Corps where he served as Radio Operator (Crypto), Truck Driver, Forward Observer, NBC (Nuclear, Biological, Chemical) Team Leader. This prepared him to become the proud owner of both dogs and cats.

He is the most interesting man in Virginia: besides working with the USDOT/FRA as a diesel mechanic, Gary has been an on-air personality at both WBEY and WVBK FM, was both DJ and manager of a saloon, and the captain of a racing sailboat. He has lived in 48 states, 8 countries – including Spain and Cuba. He speaks Pirate fluently.

He presently farms in Fairfield, Virginia, programs his friends' computers for free, and contributes to the Loka Review.

TABLE OF CONTENTS

Calculation —— — —

Discovering color - - -

Newton began developing the method of
Calculus (calculation) in 1664 when he read
recent work on optics and light by Robert
Boyle and Robert Hooke. In repeating their
observations, Newton became curious at the
refraction of light by a glass prism. Inspired
by the mathematical reasoning of René
Descartes, he investigated – and through a
series of increasingly elaborate and exact
experiments, saw mathematical patterns in
the phenomenon of color.

Editor's reference file: Scientific Method
   Science is the method by which we understand how
things work: by systematic observation, we are able to
measure variations in relationships between things
caused through controlled experiments. These
observations may be mathematically formulated to result
in a new hypothesis – which may be itself tested
through controlled experiments. When a hypothesis
fails to perform as expected, there is discovered new
variables to account for.
   While not all science is undertaken mathematically,
when it is, this is accomplished through computation
and calculation: a relationship is formulated
mathematically through algebra, and variations on its
component variables understood through calculus. When
science is undertaken without mathematics,
relationships are explored in a similar manner.

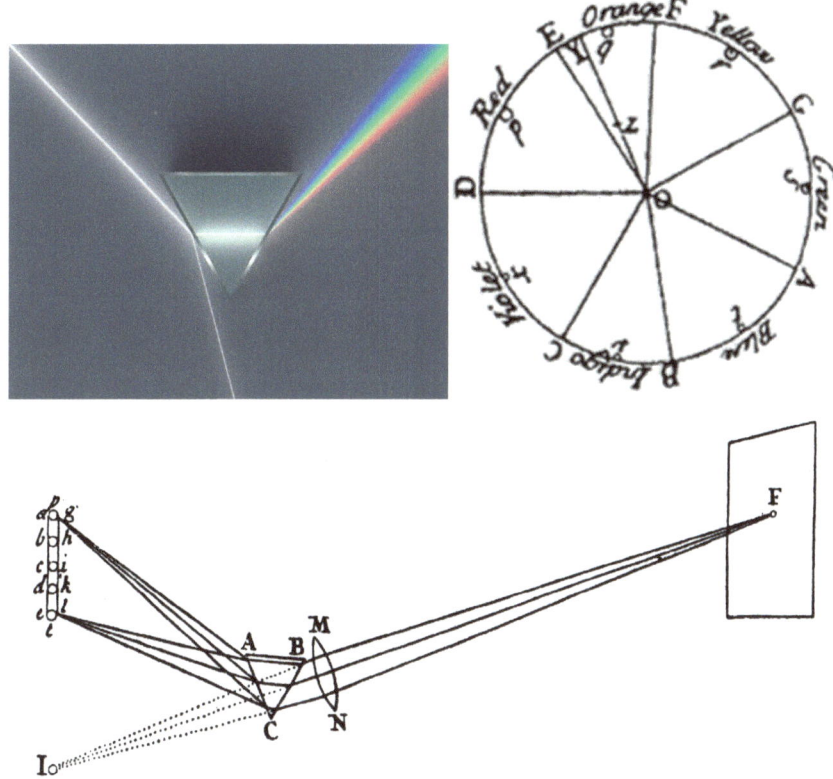

Newton discovered that white light could be "broken" into numerous colors (a "rainbow"), and then recombined into white light by the use of prisms and lenses. Measuring the composition of white light's elemental colors, Newton understood the properties of each color were different: the same white light, being bent and slowed more or less depending on whether it went through the thick or thin part of the triangular prism, resulted in a continuous pattern of color. Continuous patterns can only be understood mathematically through calculation.

Newton realized the size of each color band in the rainbow could be algebraically predicted through a continual calculation – if the angle through which it was refracted on entering or leaving the prism was known. Soon, no matter the angle of refraction, he could anticipate the result. This concept, of holding a factor constant, and calculating the effects of variables, led to the scientific method.

Editor's reference file: Prisms

A prism is a tool to refract light into rainbows. Artists will use the colors refracted to "tune" their pigments (in a similar way to how musicians will "tune" their instruments to perfect pitches of sound). Medical doctors use prisms to bend light before it enters the eye, thereby compensating for diseases which would otherwise cause blindness or impaired vision. Prisms can be used in telescopes, binoculars, and architecture for redirecting light. Scientists, such as Newton, use prisms to study the elemental components of light, and to use some of these elemental components to study other materials.

Calculate is a word that means "accounted," and describes how calculations have already accounted for every possible outcome arising from potential variations in elements of an algebraic theory.

The word itself was originally used to describe the act of having accurately counted pebbles (such as were used on an abacus). An "abacus" was an ancient tool which facilitated coded decimal counting. Pebbles (calculi) were threaded along strings like beads. For example, to count "4," 4 pebbles were moved along the first string. To count 6, 1 pebble was moved on the second string, and 1 pebble on the first: the second string counted "fives," and the first "ones." A third string counted "tens," a fourth "fifties" ... and so forth. Dozens of strings were required for very large numbers.

Because the abacus facilitated counting, it permitted simple addition and subtraction, and series of abacuses (or technique in use of existing strings) could be used to undertake multiplication or division. Modern "calculators" undertake similar procedures, but utilize self-counting electric circuits instead of pebbles.

Calculation is different than this simple counting: calculation is undertaken through algorithms which anticipate the outcome, rather than measure it. Series of additions and subtractions can be systematically organized through algebra. These systems contain elements (the series, and their required additions and subtractions). Understanding how each system's elements interact permits the result to be calculated, or anticipated.

Calculus - - -

Newton kept his method of mathematics (known as Calculus) hidden from all but his closest friends until 1704, when he published a book on it he titled "Opticks." Today, we spell the word "Optics" (no "k"): over time, written English language has changed, but the words are still spoken the same way. What do you think Newton would make of our modern use of emojis and emoticons? ☺

Editor's reference file: Science

"Calculation," or "calculus," is the process of mathematics which enables continuous computation.

"Computation" is a system of logic which permits the truthfulness of assumptions to be tested by comparing the relationship of known facts to uncertain facts under varying (variable) conditions. Computation was invented by Muḥammad ibn Musa al-Khwarizmi, and forms the foundation of modern philosophical reasoning.

Continuous computation through calculation permits an observation of the interactions of elemental components which formulate the computations, even as they varied (in algebra, these elements are known as "variables" and "constants").

In short, Newton's method of calculus permitted modern scientific method, which relies on experimentation to control the variation and constancy of these algebraic elements.

Newton was the first scientist. Though his predecessors and teachers studied natural phenomenon through al-Khwarizmi's algebraic philosophy, Newtonian Calculation permitted people to understand how these things worked.

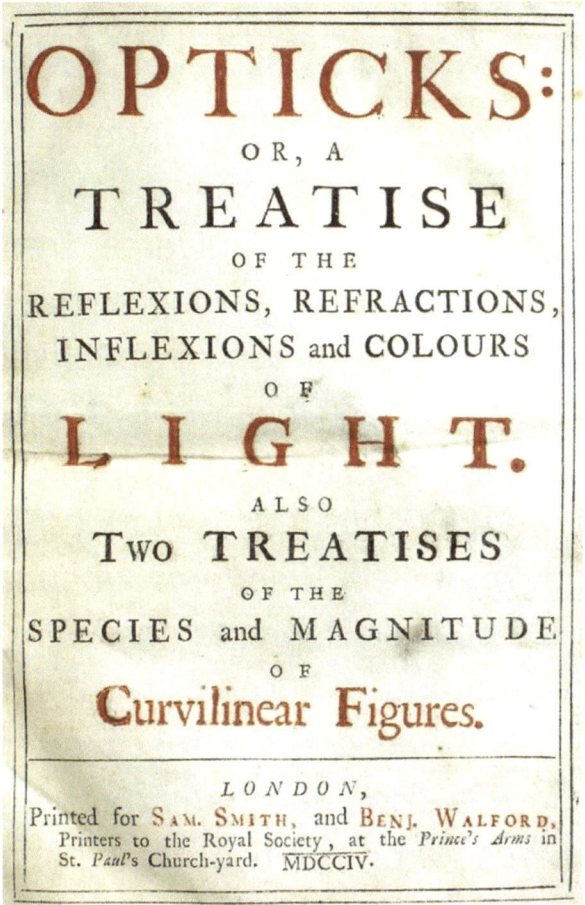

OPTICKS:

OR, A

TREATISE

OF THE

REFLEXIONS, REFRACTIONS, INFLEXIONS and COLOURS

OF

LIGHT.

ALSO

Two TREATISES

OF THE

SPECIES and MAGNITUDE

OF

Curvilinear Figures.

LONDON,

Printed for SAM. SMITH, and BENJ. WALFORD, Printers to the Royal Society, at the *Prince's Arms* in St. *Paul's* Church-yard. MDCCIV.

This is the cover page. Notice the MDCCIV? They are Roman numerals indicating the date of publication: the Romans used abacuses to count, and understood all numbers were actually units: ones, fives, tens, fifties, hundreds, and so forth. Here we see an M (1000's), a D (a 500), 2 C's (100's), an I (a 1), and a V (5). Whenever a smaller Roman numeral comes before a larger one, it is just like on an abacus: subtract the smaller number from the group: V = 5, I = 1, VI = 6, IV = 4. The date reads: 1000+500+200+(-1+5) = 1704. Newton – and many others – sometimes even wrote in Latin because most people spoke the language: just like how, today, most scientific work is written in English.

Newton is known not only for understanding light, but for discovering gravity – and the laws which govern force in the Universe, creating the science of Physics. He also created the science of History, beginning by publishing an edition of *Geographia Generalis* by the German geographer Varenius in 1672, and, after he died, his friends published his work on *The Chronology of Ancient Kingdoms Amended* (1728). He also tried applying the scientific method to theology - *Observations upon the Prophecies of Daniel and the Apocalypse of St John* (1733) was also published after he died, and is famous for his many other achievements in science and mathematics, and his numerous inventions, one of his greatest achievements is often overlooked...

...The doggy door.

The Doggy Door —— — —
Experimental optics led to the pet door - - -

Newton couldn't have made such an impact on the world of mathematics through advances in optics if he hadn't been experimenting. Coincidentally, it was his experiments in optics which led to him having an impact on the pet world as well.

Editor's reference file: Paying for College
Isaac Newton earned scholarship in the form of work-study. To pay for his early education at University, he was a waiter and a butler.

Isaac Newton
Second Law of Motion

Isaac Newton
Second Law of Motion

The story goes that during these experiments, Newton's cats kept bothering him, wanting to come in and out - as felines are apt to do. At some point, he became irritated sufficiently and altered his door so the cats could come and go as they pleased. The invention of the cat door quickly led to the invention of the doggy door.

The Coat of Arms of Sir Isaac Newton

What Sir Newton's Coat of Arms means - - -

In recognition for his important work, Isaac Newton was Knighted by Queen Ann of England on April 16, 1705, in a ceremony conducted at Trinity College, Cambridge. At the time, that honor was usually granted to military officers and senior figures in the national and local governments, as well as to rich merchants and others with political connections. Newton made this coat of arms for himself.

Alejandro Jenkins of the University of Costa Rica (2014: Isaac Newton's sinister heraldry, published by Cornell University: http://arxiv.org/abs/1310.7494) studied the origins and meaning of this unusual choice and, investigating Newton's personal writings, discovered Newton was secretly a Humanist (which is a kind of non-dogmatic practice of theism which aims to improve human affairs as a form of worship). Humanism arose in reaction to the English Christian Theocracy of the 16th Century, a period of time when religious law was used to justify genocide, enslavement, despotism – and even the execution of the King and Parliament.

The left-dominance chosen by Newton suggest through symbolic signs that he was a bastard (though "bastard" means without a father, for knights like Newton, it means without inheritance, or suggesting a new lineage – only men were able to own property in ancient England).

This may speak to his belief that his scientific revolution would begin a new era. But perhaps it is also more personal: Jenkins discovered Newton's father died before he was born, and Newton, having been born on Christmas Day without a father, was quickly abandoned by his mother to be raised an orphan before later being re-adopted by her. The choice of bones is also an important symbol. Newton looked into his family history and found a distant ancestor carried the symbol of bones during the Crusades. But Newton, may have also intended them to represent a central belief in humanism: that all humans are equal in death.

But did he invent the doggy door? - - -

Yet there remains no direct evidence that Newton actually invented the pet door – only stories. Did he, or didn't he?

Newton's method of science can help us understand not only whether this story is true, but if it is false, the reason why it came to be understood as true. But this mystery will not be solved by numbers: we require a different form of science, "History."

Science without Math ——— — —

Historical investigation - - -

Newton created history as a kind of non-mathematical science. But even though no mathematics is required, history uses the same method as mathematical sciences do. To practice historical investigation requires we compare the unknown fact to known facts, just as in any scientific investigation.

What is History?

"History" is a word that means "knowledge acquired by investigation." Whereas mathematical science uses numbers in an equation to ascertain the whether a belief is true, history utilizes a narrative. Both function upon the logical premise that if the relationship between two true facts is understood, their mutual relationship to a third uncertain fact can be discovered.

Mathematically, this is expressed as: if $a=b$, and $b=c$, then $a=c$. Historically, this would be represented as $a+b+c = 3a = 3b = 3c = 2a+b = 2a+c = 2b+a = 2b+c = 2c+a = 2c+b$... and so forth. Historical analysis is more lengthy and time consuming, but enables the application of science to things which mathematics cannot understand.

In practice, a historian will take facts which were known to have happened and attempt to understand whether a third fact did happen. For example, if it is known that there were cookies in a jar, and now they are gone, and it is also known that cookies cannot leave the jar without someone taking them from that jar, a historian would acquire the knowledge that the cookies were taken from the jar. Further investigation could result in the discovery of who took the cookies.

The following statement is true – but is this proof that Newton invented the pet door?

- *No mention is made by Newton in any of his correspondences of any pets*

By analyzing this, we see it is not evidence that Newton either had cats or did not have cats: while it is true, not everyone writes about their pets. Maybe he just didn't write about his pets!

In science, the absence of evidence is not proof that there is no evidence. Mathematically, this is expressed as $1 + x = 0$. Just because 1 does not equal 0, does not mean that there is not a fact (a number) which, when combined with one, would equal zero (in this case, that number is negative 1).

We need proof – one way or another – that Newton either had cats or did not. The following is also true – is it a fact?

- *At a meeting of the Royal Society on 24th December 1719, Newton mentioned his dog in passing: she had recently gone blind with cataracts*

This is indeed a fact – one which proves Newton had dogs. But it is not a fact which proves Newton had cats. Mathematically, we would understand this as $1 + x = y$. Without further information, the values of x or y cannot be known for sure, even if we do know that x is 1 larger than y (in this case, we do know Newton kept a dog, but nothing certain about whether he kept a cat).

Let's continue our investigations…

Newton's dog, named "Diamond," almost prevented him from completing his most important work. Some remember how one night Diamond got so excited that there was a visitor at the door that, jumping about, knocked over a candle, igniting important papers Newton was working on (his theory of gravity). The work was a total loss. But Newton was a gentle man - he had already understood that dogs and people see the same world in the same light, but understand what they see very differently. He knew the dog did not intend to cause such harm. So he simply lifted up his dog and gave it a hug, saying, "Oh, Diamond, Diamond, little do you know the mischief you have done me!" It would take Newton a full year to rewrite his theory on gravity.

What about these facts? Do they prove Newton had cats?

- A biographer of Newton, Richard S. Westfall, in his book "Never at Rest" said Newton was a vegetarian because he loved animals, and could not tolerate the necessary cruelty required to kill them.

- Voltaire, a contemporary of Newton, said "Newton had cultivated this sentiment of humanity, and he extended it to lower animals. With Locke he was strongly convinced that God has given to them a proportion of ideas, and the same feelings which he has to us. He could not believe that God, who has made nothing in vain, would have given to them organs of feeling in order that they might have no feeling. He thought it a very frightful inconsistency to believe that animals feel and at the same time to cause them to suffer. On this point his morality was in accord with is philosophy. He yielded but only with repugnance to the barbarous custom of supporting ourselves upon the blood and flesh of beings like ourselves, whom we caress, and he never permitted in his own house the putting them to death by slow and exquisite modes of killing for the sake of making the food more delicious. This compassion, which he felt for no other animals, culminated in true charity for men. In truth, without humanity, a virtue which comprehends all virtues, the name of scientist would be little deserved."

No, while it demonstrates Newton liked animals, and had a dog, all we know is if he had cats, he would have loved them.

What about this?

- *Newton was known to have disliked pets in the home, believing them to be dirty and troublesome*

This seems to contradict what we understood so far and may actually be not true: Newton certainly kept a dog! Who says this, and why? Do they have proof of this statement? It contradicts all the facts we have so far, and there is no reason to believe this statement: in science we would say this is not true. Mathematically, this would be like saying, 1+2=0. It doesn't add up.

We still need proof that Newton had cats.

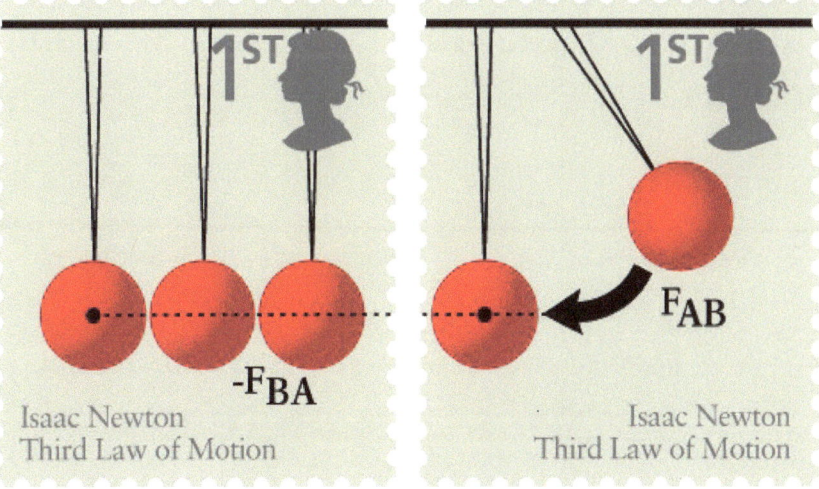

Is this fact evidence that Newton had cats?

- *John Maynard Keynes, the influential economist, who bought a large number of Newton's non-scientific papers, especially on alchemy, said in his memoirs that Newton remarked his cats were growing fat on Newton's uneaten food (Newton worked so hard that he ignored the meals brought to him by friends concerned he was not eating)*

This fact requires we take Mr. Keynes word for what those letters say – and whether we agree or disagree with his conclusions of economics, he is a scientist who is trustworthy in reporting facts truthfully. This would be proof Newton had a cat. We can compare this truth to the unknown truthfulness of the story of the cat door, and conclude:

1. *Newton had cats*
2. *Newton liked animals enough to be responsive to their needs*
3. *It is likely that Newton would have damaged his door to permit his cats to come and go as they liked*

However, other scientists practicing history, specifically the historians S. Brodetsky, Louis Trenchard More, and Alfred Rupert Hall say that people who knew Newton said "Newton kept neither dog nor cat in his chamber" (chamber is a fancy word for "room").

But maybe Newton simply didn't own pets during that time? Some people will own a dog or cat, and then not own a dog or cat.

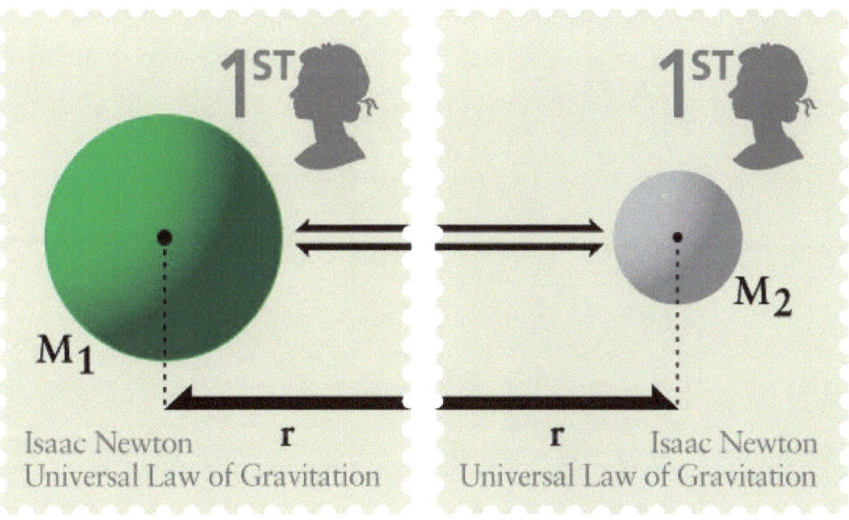

So, what do you think? - - -

Did the inventor of calculus, the first scientist, the man who discovered gravity and the fundamental laws of physics, and accomplished so many other great things, also invent the doggy door?

The lesson of science —— — —

Science teaches us that sometimes, we don't know for sure whether something is true or not.

Newton saw this in mathematics, where the idea that the calculation of 1/x may come close to equaling zero – but never equal zero – seems to counter every day experiences. We are unused to either mathematical or practical limits to our understanding.

Newton also saw this in history, and non-mathematical science. We like to think things are very certain – true or false. Yet we have seen this in trying to understand if Newton invented the doggy door.

We cannot say Newton invented the doggy door, but we also cannot say that he did not invent the doggy door. We may only say it seems very likely he did. This "theory" is as reliable as Newton's theory of gravity.

Other theories form the foundation of our modern understanding: we must function upon premises of belief, whether those beliefs are in a theory of gravity, a theory of subatomic particles, Darwin's theory of evolution, or our own theory that Newton invented the doggy door.

And sometimes, as technology improves, we are able to get definite proof: the theory of bacteria was confirmed when technology improved so that Antonie van Leeuwenhoek could actually see microscopic organisms.

Newton recognized some calculations can never be computed, and that some things cannot ever be known for sure. The "limit" of understanding requires a reasonable scientist to develop confidence in their beliefs: even when we do not have positive proof and all the facts, it is possible to still make a logical conclusion...

...That Newton invented the doggy door.

www.ingramcontent.com/pod-product-compliance
Lightning Source LLC
Chambersburg PA
CBHW040820200526
45159CB00024B/3055